# Where is God?

A Childish Question

**Redha AL Yousef**

**1 |** A Childish question

# Contents

Introduction .................................................................................. 6
First Chapter: ............................................................................. 10
   Scientific Fact (Earth is Protected): ..................................... 12
   The Aspect of Argumentation: ............................................ 12
   The Difference between the Old (Eternal) and the Recent (Created) : ................................................................................ 20
   A Benefit: ............................................................................. 20
Second Chapter: ......................................................................... 24
   In Knowing God .................................................................... 26
   The Oneness of God : ........................................................... 32
Third Chapter ............................................................................. 38
   Attributes and Names of the Creator Lord .......................... 40
   Summary: ............................................................................. 41
Fourth Chapter ........................................................................... 44
   Cosmic Gravity Waves: ........................................................ 46
   The Big Bang: ...................................................................... 47
   The Speed of Light : ............................................................ 54
   Jupiter: Cosmic Guardian .................................................... 61
Fifth Chapter .............................................................................. 66
   Is Empty Space (Nothingness) Present in the Universe?: ... 70
   Entropy and Life: ................................................................. 72
   Negative Entropy ................................................................. 73
   Gibbs Free Energy and Biological Evolution ...................... 75

Entropy and the Cosmic System: ........................................................77

How does the cosmic system form?: ................................................80

Conclusion................................................................................................94

## Introduction:

Indeed, God is not to be limited to a place or a state. How can you confine the Creator of time and space to what He created? How can a place take Him in when He is the one who originated it, fashioned it, established it, and set in motion the causes and laws by which it operates toward perfection? God is the Creator, the Predestined, and the Causer of its course, and He safeguards it from disappearing. Through this book, you will reach the innate knowledge of the divine existence and the certainty of the soundness of the principles of religion, servitude, and monotheism, from a smooth scientific perspective. It would be easy for the noble reader to understand the essentials of monotheism and faith in the divine existence, without extremism, racism, or adherence to a specific religion, sect, or ideology, through all that it contains rational and scientific evidence that stimulates the innate inquiries of individuals, regardless of their religion, orientation, or age.

# 5 | A Childish question

# Chapter one

**7 |** A Childish question

## First Chapter:
## The Truth that God Exists

Does God really exist and how can you prove it with your mind? If you spread your sight throughout this great existence, you shall see parts that are connected and rely on each other. Nothing stands unless another thing is completing it and supporting its functions, just like the buildings that need each other so much that if any part of it is missing, it cannot function on its own and cannot be complete and cohesive except through the interconnection of its parts. Let's take, for example, to bring our minds closer to understanding this. You may say that no matter how grand and changed the plant may be, its beginning always goes back to a tiny seed. Its cells divide to build itself in a specific coordination and a fixed geometric algorithm. It grows in a system that cannot be disturbed or changed on its own. It is not self-sustaining, but rather a part of a magnificent interdependent system, interconnected with complementary functions, forming a grand structure called existence. This plant relies on water, air, light, soil, and weather. It is influenced by them, and it, in turn, influences them. They support each other to survive through successive life cycles, maintaining a sustainable rhythm. This complex system, with its intricate details, begins with the formation and division of cells, shaping the plant's parts. How did the root cells know they would become roots?

**Similarly, the stem cells, leaf cells, fruits, and seeds. Once again, there must be a regulator and guide that does not err or tire, unaffected by any change that may occur. It created them and established their system and processes. The plant is in constant need of this regulator. If it were absent for a moment, it would wither and disappear, unable to sustain itself. In short, there can be no construction without a builder, no action without an actor, and no composition without a component that forms it.**

## Scientific Fact (Earth is Protected):

Despite the many comets and asteroids that have passed through our solar system over the ages, Earth has only been directly hit by a few ancient impacts from thousands or even millions of years ago. This is because of the amazing arrangement of planets in our solar system, where Jupiter's strong gravitational pull and large size absorb all comets and asteroids heading directly toward Earth. This is just one example of how the planets and their magnetic fields work together to protect our planet, with other roles still unknown. All of these factors contribute to the safety and longevity of Earth.

## The Aspect of Argumentation:

If you see a building, you can be certain without any doubt that there must have been a builder who constructed it, even if you did not see them. Similarly, if you look at the horizons and within yourself and see a well-designed creation with such precision and order, you can infer the existence of the Creator. The Creator of senses has given them a heart to which they return, and He has made the senses indicate the apparent, through which they infer the Creator. Thus, when the eye sees a connected creation, the heart is guided by what it observes, and when the eye guides the heart to what it observes, the heart reflects on what it has seen. From the kingdom of the heavens and its unfathomable elevation without pillars that hold it, it never declines nor advances. It does not descend to draw near nor ascend to withdraw. It does not change over time or differs with the changing nights and days. It does not incline in any direction, nor does any part of it collapse. Likewise, the eyewitnesses the constantly moving stars, each following its own orbit, shifting within the constellations day after day, month after month, and year after year. Some move quickly, some move slowly, and some move moderately. They return, align, take positions, lengthen, and shorten according to the sun, which supervises their appearance when it sets, and the sun and moon continue their celestial journey within the constellations, steadfast, unchanging in their times and seasons. Those who understand the calculation of a known subject know that this is not from the wisdom of man or the search for illusions or the turning of thought. Thus, when the eye guides the heart to what it has witnessed, it recognizes that the One who holds the heavens from above is the Creator of the heavens. Then the eye looks at what is beneath it on the Earth, and the heart is guided by what it observes. The heart, with its intellect, knows that the One who holds the leveled Earth from falling or descending, while it is like a feather in this vast universe, is the same One who throws it and it falls in its place, as it is light in weight. So, who is it that holds the sky above it?

If it were not for that, it would collapse with all that is upon it, with the weight of mountains, creatures, trees, seas, and sands. The heart recognizes by the guidance of the eye that the One who manages the Earth is the Master of the heavens. Then the ear hears the sound of the strong, stormy winds, and the gentle, pleasant breezes. The eye observes what is uprooted from the bones of trees, what is demolished from strong structures, and how the weight of sand shifts from one place to another without any prior notice neither seen by the eye nor heard by the ear, nor perceived by any senses. It is not tangible to touch, nor limited to observation. Then, the pen increases the eye, ear, and all the senses in affirming that there is a Maker. This is because the heart contemplates with the intellect within it, and it knows that the wind does not move by itself. If it were the one in motion, it would not cease moving, nor would it demolish one group and miss another, or uproot one tree and leave another beside it, nor would it strike the earth and veer away from another part. When the heart contemplates this matter, it knows that there is a mover who directs it where He wills, who resides within it as He pleases, and who affects whom He chooses.

Likewise, the eye and ear guide the heart in realizing this grandeur, and the heart knows it without the need for other senses. When the senses indicate the movement of this magnificent creation from the Earth, its density, transportation, length, width, and the weight of mountains, waters, creatures, and others, and it moves in one direction without moving in another, while it is a unified body, a connected creation without division or separation, it collapses in one area, subside in another, and rises in yet another. At that moment, the heart recognizes that the Mover of what has been moved is the Holder of what has been held. He is the Mover of the wind and the Holder of it. He is the One who manages the heavens, the Earth, and what is between them. If it were the Earth itself that caused the shaking, it would not shake, nor would it move. But it is He who has orchestrated and created it, causing it to move as He wills.

Then the eye looked at the magnificent signs of the cloud that is suspended between the sky and the Earth, resembling smoke with no tangible body touching the ground. It passes through trees without moving anything from them, nor does it produce any sound or leave any trace. It intercepts travelers, causing darkness and density, carrying the weight of water and its abundance beyond imagination. It carries thunderbolts, shining lightning, thunder, snow, hail, and ice, exceeding the limits of perception and defying the comprehension of the heart. Minds cannot grasp its essence, nor can hearts comprehend its wonders. It moves independently in the air, gathering after dispersal and merging after separation. The winds blow from all directions, directed by the command of their Lord, sometimes subsiding and other times ascending, carrying abundant water. When it releases its water, it disappears, disperses, and goes where it cannot be seen or known. The eye conveyed this to the heart, and the heart understood that if this cloud were not orchestrated by a planner and if it possessed all these qualities spontaneously, it could not bear even half of the weight of the water it carries. If it were the one sending forth the water, it would not be able to bear the burden of thousands of leagues or more. It would have sent it closer and in greater quantity. It would have destroyed structures, damaged vegetation, and arrived in some places while leaving others untouched. The enlightened signs indicate to the heart that the Manager of affairs is One, and if there were two or three, there would be differences in their management, contradictions in their actions, delay in some, and advancement in others. Some would subside what has ascended, and others would raise what has subsided. Something would appear and then disappear, either being delayed or preceded by something else. The heart recognizes that the Manager of all things is the One who is not absent from anything and everything that appears or disappears. He is God, the First, the Creator of the heavens, the Upholder of them, the Spreader of the Earth, the Manager of what lies between them, and much more that we have enumerated and beyond what we can enumerate. The eye also observed the alternation of night and day, two new and perpetual phenomena that do not cease in their duration and do not change in their frequent alternation. They do not deviate from their state, with the day in its light and brightness, and the night in its darkness and shadow. One enters into the other until each reaches a predetermined and known limit in terms of length and shortness, in a single rank and a single course, with the stillness of those who dwell in the night and the movement of those who move in the day, and the movement of those who move in the night and the stillness of those who dwell in the day. Then there is the heat and the cold, with one following the other promptly, so that heat becomes cold and cold becomes hot in their appointed times and manifestations. All of this is evidence by which the heart is guided to the Lord, glorified and exalted be He. The heart, with its intellect, understands that the One who has arranged these matters is the One, the Almighty, the Wise, who has always been and always will be. If there were Gods with Him in the heavens and the Earth, each God would take away what it created, and some would prevail over others, causing corruption between them. If it becomes clear to you, upon reflection and observation, that all beings and entities require a Creator and Maker, then you should inquire about the identity of this Creator and Originator.

The author can refer to the tool or its owner, like saying "The rope was cut" or "It was cut by the knife," and when you say "I wrote the book" or "My pen wrote it," both statements are correct logically and conventionally. The origin of action and movement becomes evident through the instrument. Thus, the Author, if desired, becomes multiple with the multiplicity of effects, and here the instruments are called causes. All events in this world are caused, for writing occurred due to ink and the movement of the pen, and the pen moves due to your hand's movement, and your hand moves by your own motion. And so, for example, the wave moves by the movement of air for the water, and the air moves the rising vapor due to the motion of the sun's rays, and the sun's rays move by the motion of the sun. Therefore, there is no doubt about the multiplicity of causes and instruments and their existence on top of one another. However, all these instruments are associated events that must have a Creator who is a driving force with willpower and wisdom to manage these actions without association or movement. If it were otherwise, he would also be a machine and a cause; rather, he must be above them all. Beyond these instruments, the Possessor of the instrument is the Creator, the Maker, and the Source of actions. It must be uncreated, not brought into existence by anyone else, independent from anything other than itself. If it were a created being, it would be in a state of need, dependent on other causes and instruments. Above it would be an eternal, ancient, rich Lord who governs its affairs and actions. The ancient, eternal, and rich Lord is the ultimate goal and the end of all ends. He is the mover, the director, the creator, and the initiator of everything. He is rich beyond comparison, while everything else requires Him. He is the one who sent messengers to His creation and revealed books to them, so that they may know their purpose and the ultimate goal of their creation. He informs them of what He desires from them. Those who disobey Him, He is God.

## The Difference between the Old (Eternal) and the Recent (Created) :

A thing could either be old or recent, they do not gather nor their contradiction could ever be lifted. What is recent could never be new and the recent new thing could never be old. There is no doubt that the recent in all its meanings requires an Updater in the impoverished terminology, which is dependent on others and does not stand on its own. This "other" is the creator who doesn't rely on anyone else. It is timeless, eternal, and abundant in ways that surpass everything else. All the changing and overlapping elements in this world, and their interdependence, are evidence of the influential creator, the ancient maker.

## A Benefit:

Everyone, when they reflect on themselves, finds a need to seek refuge in a strong and secure shelter. This need is not exclusive to kings, scholars, or wise men; everyone experiences it. We all need someone to turn to during difficult times, someone who can teach us when we are ignorant and strengthen us when we are weak. This is why every soul seeks shade under something greater than itself, something that can protect it from harm. It is an instinct, like hunger or thirst. Human beings often call upon someone greater than themselves to seek shelter under their shadow and take refuge in their protection. However, if this instinct is corrupted, one may seek refuge in things that are harmful or inappropriate. They may turn to the weak and needy, just like themselves. Their requests may revolve around habits, instincts, desires, waywardness, and mischief. For instance, their inclinations may lead them to crave eating clay and coal instead of nourishing food that benefits the body, which is consumed by those with balanced temperaments. Those who ascend from the land of distortion and deviation to the sky of moderation will certainly moderate their requests. They seek self-worth without degradation, richness without poverty, strength without weakness, knowledge without ignorance, completeness without deficiency, nobility without blame, generosity without reproach, and integrity without compromise. Hence, seeking any other refuge or relying on it is inappropriate. How can a needy person ask another needy person? How can a humiliated person seek empowerment from someone equally powerless? How can a contented soul accept asking someone other than themselves and not seek assistance from the aforementioned "other"? And how can you be satisfied with seeking refuge in someone who seeks refuge in others and doesn't seek refuge in themselves? Indeed, seeking refuge in the most secure and relying on the most trustworthy and complete is deeply ingrained in every nature. Seeking the assistance of the sublime and magnificent Lord, the Creator, Provider, Mighty, and Wealthy is a natural and necessary inclination. The pursuit of knowledge about Him is a rational obligation before a religious one because there is no religious obligation before acquiring knowledge. This connection indeed has levels. We cannot connect with the Lord except through the prophets, and we cannot connect with the prophets except by following their guidance.

We cannot connect with them except through steadfast scholars, and we cannot connect with them except through supportive brethren. The intellect necessitates all of this when it realizes its own need for them. It is a unified connection with various levels that are interconnected, as some levels complement others. In general, the intellect does not permit seeking anything less than perfection at each level, and nothing is perfect at every level except these mentioned ones.

To bring the meaning closer, let's say you were on board an airplane or a ship and it broke down in a place where neither a ship nor an airplane could rescue you. Would your heart cling to the hope that something or someone could save you from your problem? That which you seek for salvation is God Himself, who is capable of rescuing and delivering you. Through this, the existence of God is affirmed

# A Childish question

A Childish question

# Chapter two

# A Childish question

## Second Chapter:

## In Knowing God

Here, we need to discuss the concept of knowledge and its limitations when it comes to understanding God. Due to our weaknesses and limited tools of perception such as sight, hearing, touch, and imagination, we can only attain a limited amount of knowledge about God. Throughout history, the relationship between creation and the Creator has been based on worship, which is a branch of knowledge. We cannot worship what we do not know, and therefore, knowledge of God is the foundation of this relationship.

One of the most important sources of knowledge about God is what He has revealed about Himself through the prophets and their scriptures, which we accept and believe in.

Believing that God is One because multiplicity corrupts creation and existence; there is no likeness to Him because comparison leads to association which leads to multiplicity and similarity that negates independence and richness while generating neediness and composition. We recognize Him through His effects and signs that indicate His power; we do not know Him through Himself but rather through evidence that points toward Him.

Association with the senses ends with knowledge of something tangible that is not God because what can be sensed or touched or seen is limited and created by a creature. God cannot be comprehended by our senses or delusions alone; they are incapable of encompassing Him. Instead, we infer His greatness from His creation's craftsmanship and design.

What is the path that God made to know Him? God Himself is not perceived by a sense, nor is He measured by anything. The first path to knowing God is to think about His effects, His creation, and the reason for its existence, what is seen from it and what is not seen, what is surrounded and what is not surrounded, what is heard and what is not heard, starting with yourself.

(We will show them Our signs in all the horizons and themselves until it is clear to them that it is the truth)

Let me provide you with an example to better understand the meaning. Imagine describing a house. You would mention its structure, rooms, windows, doors, and roof. You describe everything based on what it actually is. On the other hand, if you describe a barren land, you would say there are no buildings, houses, or any signs of human construction. You negate all forms of construction because you haven't observed them, not because you are unaware of the land's existence. Your description is based on what you have seen with your own eyes, the truth you know. Similarly, God has described Himself as being beyond human perception. He cannot be seen with physical eyes, comprehended by human thoughts, or represented by mere conjectures. All the attributes attributed to God in various accounts emphasize His transcendence. Reflect upon this: the apparent forms are expressions of psychological images, which are embodiments of mental concepts known as truths.

Everything that you are informed about is based on what the informer has personally experienced, witnessed, and comprehended. Otherwise, they would be unable to express it. Similarly, everything that reaches your ears is within your realm of understanding, as it belongs to your own kind. It descends to you from where you are, and it does not ascend beyond your capacity. Whenever you speak or are spoken to, it engages you in its different levels, stages, realities, mental images, words, meanings, intentions, and verifications. But anything beyond your reach cannot be approached, intended, or desired. You cannot indicate it through presence or absence, gestures, hidden meanings, or subtle expressions. Even the most precise mental indications cannot refer to what is beyond your nature or comprehend what is not of your kind. One with a limited perspective cannot direct themselves towards the realm that transcends perspectives, and one with a certain rank cannot aim for or desire what is beyond that rank, even with the utmost intention. In essence, one is who they are, independent of knowing their own self. As for their creation, it returns from description to description.

However, even though God's essence is beyond the rasp of our senses, He has described Himself to you in a way that resonates with what you say, and you find confirmation of all of this within yourself. The purpose of this is for you to understand what you say and recognize the true speech, for it is the ancient self that knows nothing but itself. As for everything else, all that is said about it consists of attributes, qualities, manifestations, and praises that lie between the unseen and the witnessed. On the other hand, you can hardly know someone except through how you describe them. If asked about someone, "Do you know them?" and you respond, "Yes, I know them. They are tall or short, wide or thin, white or black. They are intelligent and capable," and so on without describing their essence, it is clear that your knowledge of them is limited to their names and attributes. Understanding oneself in appearance is possible through knowledge of one's attributes, but true self-knowledge is impossible. Knowledge is limited to names and attributes, and the reference point for these attributes is other attributes rather than the essence itself.

Therefore, it can be concluded that knowing God cannot be like knowing other creatures because God does not appear through sensory perceptions such as eyesight or hearing or other visible or imaginary perceptions such as emotions, knowledge, reason, etc. If He were to be perceived by these means He would be of the same nature as them and would be created rather than Creator. It is evident that the purpose of knowing God and worshipping Him as commanded in all religions must be possible for us to achieve. Since self-knowledge of God's essence is impossible for the reasons we have outlined, certainly, what is meant by knowing God in the context of His creation and His signs is what has been made possible for His creation.

Undoubtedly, when a person sees the signs of God and recognizes their greatness in every situation, they become submissive and humble. This is the meaning of worship which God has commanded us to do. He did not command us to worship His invisible essence but rather He has placed levels and signs behind Him which are manifestations of His attributes. He has commanded all creation to observe these levels and signs so that they may learn from them, emulate them and become similar to them because God loves His attributes and loves those who embody them. This is the meaning of worship; obeying God by embodying divine attributes.

## The Oneness of God :

The desired oneness in the ancient self is the simple indivisible oneness that cannot be accompanied by anything else. It encompasses existence, essence, possibility, affirmation, negation, reality, and consideration. All of these are external factual realities. Anything that lacks reality has no consideration. As for negation, it is a branch of affirmation, and it is something that cannot be expressed. The denied is a present term for a nonexistent meaning, conceived by incomplete understanding as if it ascribes existence to non-existence spontaneously. The emptiness of a house is something, but it does not imply the presence of a person, therefore, negation is something. As for possibility and the universe, the original meanings exist externally for verification. Otherwise, there is no specialization for meaning with anything. Goodness, then, becomes a form of evil, and victory becomes betrayal.

If there is something that applies only to it, it becomes an accident. And as for the ancient, nothing is imposed upon it.

It encompasses, through its novelty, all assumptions, contingencies, directions, considerations, essences, attributes, limitations, boundaries, indications, possibilities, actions, effects, and changes, to the point where mentioning any of them is denied in its essence. All of them have a dark appearance by their command, and what does not apply to it does not apply to it, nor does what is initiated in it ever return. The intention behind not dividing it is not its smallness that prevents division, but rather its vastness that encompasses everything else in knowledge and power, even in the meaning of non-being. Having a description with it becomes a qualifying attribute, and the inclusion of every description under it becomes an implication.

In terms of the concept of truth, if you reflect, for example, on the term "sun" and its application to the celestial disk or the emitted ray, you find that in common usage; both of them represent truth due to the presence of signs of truth in both of them, such as spontaneity, the absence of negation, regularity, and others. It is not due to the need for evidence in one of them over the other, nor is it based on transmission and addition. Generally, it is not due to verbal or conceptual commonality because there is no commonality between the substance of the ray and its image. They are two distinct manifestations resulting from the disk and the inclination of the ray, similar to how your reflection in the mirror is separated from you. Its substance is not the same as your body, and its image is not of the same nature as the image of your face. It is a shadow image reflected from the image of your face, thus it is not based on improvisation.

The term "existence" is often used to refer to God and the truth, but those who discuss this topic do not fully understand the different ways in which this term can be used. Some argue that it can be used both literally and figuratively, while others argue for a verbal or conceptual unity between them. However, the truth is that the existence of God is simple and cannot be compared to anything else. It is important to note that our limited perception can only lead us to understand what is created, while God is the Creator who cannot be fully understood by us. Therefore, we can only know God through His effects and signs. It is also important to recognize that belief in God is singular and that multiplicity and comparison are not applicable to Him. The use of the term "existence" can be applied to both the truth and creation, but it is important to note that the primary use of this term is for the Creator, as everything else is a result of His actions. Overall, all words in the world have meanings related to this concept because everything that exists is a result of the Creator's actions. Thus, regarding the term "existence" and its application to God and truth, the proponents of discourse and statements on this subject hardly understand the different uses of this term. As a result, they have fallen into blind darkness. Sometimes they say that the application of existence to God and creation is based on both reality and metaphor, while other times they argue for verbal commonality or conceptual commonality. They also argue for their union, and there is nothing but true existence and restricted existence. However, the truth of the matter is that the existence of the Truth, exalted be He, is simple and not inclusive of anything else. Therefore, the application is necessarily based on conceptual commonality. It does not mean that when used, it excludes the universe outside. The usage is not without purpose, as it perceives in each case what illuminates non-existence.

And for that reason, it is said to be existent, even if this aspect is not observed, it is truly existent because it is an independent being. Thus, its application to the truth is obvious and commonly used in both truth and creation. There is no escape from saying that each of them is based on reality, and there is no commonality between them. Thus, we must acknowledge that it is based on reality after reality due to their non-equivalence in existence and the fact that creation is the effect of His action, and His separate entity is like a shadow of His connected action.

The term is first applied to the effect and then to its corresponding reality, which is the observable relationship between them. Otherwise, there is no connection between it and the essence of the effect. If one were fair and just, they would find that all the words of the world carry these meanings from this perspective because every existing thing is the result of its cause, and it does not possess what the cause has given it in terms of the shadow of its connected form. The term applied to it is primarily due to its cause, and from it, it descends to its effect. All goodness's are the result of the principle of good, and all evils are the result of the principle of evil. The truth is only realized in the specific situation, as its impact is not noticed until it is unfolded. The subject becomes multifaceted due to the various manifestations and qualities that arise from it. When observing the impact on oneself and the general subject, its qualities become more apparent. The intention was for one, not for many.

In the apparent proverb, only the sultan added and only the sun rose, thus illuminating everything. It cannot be said that you are both the lamp and the light because they do not count together. Only the lamp exists in this way,
and when it comes, you are undoubtedly the light. Existence is based on truth and applies to all things without exception. As for hidden entities, there is no name or image. Therefore, a wise person must believe that God is eternal, stable, present, and the creator of this complex world composed of its parts. He is one in essence with no opposition or comparison; he cannot be described or limited by anything else - neither by

**negation nor affirmation - as everything else is dependent on him alone without any mention or formation of other**

# Chapter three

# A Childish question

# Third Chapter

## Attributes and Names of the Creator Lord

One of the qualities of God is that the first aspect of religion is knowledge of God, and the perfection of this knowledge is to believe in Him. The perfection of belief in Him is to affirm His oneness, and the perfection of His oneness is sincerity towards Him. Sincerity towards Him involves negating attributes from Him, testifying that every attribute is other than what He is described as, and testifying that what He is described as is other than an attribute. Therefore, if you see a reflection of yourself in a woman, there is nothing but you, your reflection, and the woman. Your reflection exists without you and appears to the woman; it is your visible name to her. The visible and appearance are united in external existence but differ in internal existence. This is the meaning behind saying that attributes are derived from essence; they cannot be associated with or attributed to it. They testify that they are not what He is described as, just as a reflection in a mirror does not define one's identity. Attributes are defined by their external existence when seen by others and by their internal existence when seen by oneself; these two aspects coexist like your reflection in the mirror.

## A Childish question

If you understand the positions of attributes, you will notice God's unity through His attributes because nothing appears except for Him alone. This can be seen through the absence of anything else appearing alongside Him in numerical ranks except for one (as seen with numbers 1-10). Similarly, one can see a writer's presence in books, worlds, metalsmiths, carpenters, standing people, sitting people, etc., just as one can see a lamp's light shining through all its colors without seeing any other light or sound except for its own. Describing God with attributes reveals His manifestation through them while negating them from Him reflects His abstention from them.

## Summary:

The oneness of God in His attributes has multiple levels. One of these levels is that His oneness in His attributes means that He is the only one described by them. What is seen in His attributes within creation is not a shared attribute with them; rather, it is what He created and described His creation with. For example, He is the All Knowing, and what is observed in creatures from knowledge is the effect of His knowledge and His creation manifested in them by His estimation. The same applies to all His actions. Furthermore, His attributes have appeared to the extent that they have concealed all of creation and their attributes, except for His attribute that has manifested in things.

Additionally, everything in the universe bears His attributes: the essences, essential attributes, and incidental attributes, because they are the effects of His attributes. The meaning of this is that nothing exists except His attributes and their effects, for there is no light except His light, and things are only the effects of His attributes.
Just as when you look at the sun, you find nothing but the sun and its rays, which are its effects. Names and attributes are creations, and their meaning is God, to whom differences and divisions are not befitting. Rather, divisions and combinations belong to the created, and they cannot be described as few or many. However, God is eternal in His essence because everything other than the One Lord is divided, while God is One, not divided or assumed with scarcity or abundance. Every divided entity indicates a creator for it. So, when we say that God is capable, we affirm that nothing can overpower Him, and we negate weakness from Him, placing weakness in everything other than Him. Similarly, when we say that He is All-Knowing, we negate ignorance with this word and place ignorance in everything other than Him. Thus, when God annuls things, the forms, names, and divisions are annulled. Yet, He remains the All-Knowing and the All-Powerful for those who remain. Similarly, we named

Him All-Hearing because nothing escapes His perception through hearing, not because He has the mechanisms of hearing in His head. Likewise, we named Him All-Seeing because nothing is hidden from His perception through sight, whether it be colors, forms, or anything else. This is the simplest way I have found to convey and simplify the meaning without compromising its essence. In a later chapter of this book, I will present some texts from hadiths, verses, and heavenly books that carry some of these virtues, affirming these truths.

# Chapter four

**43 |** A Childish question

# Fourth Chapter

## A Glimpse of the Miracles of the Creator

Lately, I've been exploring some fascinating scientific phenomena that have been recently discovered. I've conducted investigations and gathered valuable information about them. In this book, I will share some of these discoveries to support the topics discussed and establish key concepts that reinforce the title of this book.

## Cosmic Gravity Waves:

What are gravity waves?

Gravity waves are "ripples" in spacetime caused by some of the most violent and energetic processes in the universe. Albert Einstein predicted the existence of gravity waves in his general theory of Relativity in 1916.

Einstein's mathematics showed that massive accelerating objects, such as neutron stars or black holes orbiting each other, can distort spacetime in a way that propagates "waves" of spacetime, spreading in all directions away from the source. These cosmic ripples travel at the speed of light and carry information about their origins, as well as evidence about the nature of gravity itself.

The most powerful gravity waves are produced by catastrophic events such as the collision of black holes and supernovae explosions (the massive stars that explode at the end of their lives) and the collision of neutron stars. Other waves are expected to be generated by the spinning of non-spherical neutron stars, and perhaps even remnants of gravitational radiation caused by the Big Bang.

## The Big Bang:

Could anyone have imagined or envisioned this scientific fact just one century ago?

Who would have known about it, if it was not for the emergence of the field of physical cosmology and its significant progress in exploring the universe during the past century, acquiring tangible physical evidence of the universe's origin, and answering important questions about how this universe began?

The Quran informs us about this hidden matter and provides a clear and explicit answer to the question about the birth and origin of this universe: that the Earth and the heavens were originally one, and God separated them.

For over thirteen centuries, the human mind remained completely incapable of knowing this brilliant scientific fact. Humans stumbled into speculations, myths, and fallacies regarding the origin of this universe until knowledge, sciences, technologies, and observatories advanced to lead humanity to the truth that the heavens and the Earth were once a connected piece! Scientists recently proved this scientific fact through what is known as spectral analysis, yet the Quran preceded them by over 1400 years. It was only in the late 1920s that science concluded that the universe was once a single mass that exploded and expanded, forming galaxies, stars, planets, and other celestial bodies. Reflect on this verse: Is it not the same reality as the cosmic rupture mentioned in this verse? It is the best explanation reached by scientists regarding the origin of the universe!

*(Have not those who disbelieve known that the sky's and the earth were of one piece, then We parted them, and We made every living thing of water? Will they not then believe?)*

This verse indicates the formation of the universe and the beginning of creation, as astronomers and astrophysicists have struggled for centuries to imagine the moment of this universe's birth but failed to do so.

Instead, they arrived at the theory of the Big Bang, which is the most widely accepted theory in modern physical cosmology, based on several observed facts. The idea of this theory can be summarized as follows: at the beginning of its existence, billions of years ago, the universe was a single part in an extremely hot and dense state. It then exploded and rapidly expanded. Most of the atoms that resulted from this Big Bang were hydrogen and helium, with a small amount of lithium. Massive clouds of these primary elements gravitationally collapsed, giving rise to the galaxies and stars we observe today.

Physical calculations suggest that the size of the universe before the Big Bang was nearly zero, in a peculiar state of matter and energy density, where space and time cease to exist, and known laws of physics come to a halt. This stage is referred to as the "Ratq" (One piece). Then, this primordial entity exploded in a major phenomenon known as the Big Bang, which marks the stage of "Fataq" (Parting), transforming into a sphere of radiation and elementary particles that expanded and cooled at extremely high speeds, eventually forming the cosmic smoke from which the heavens and the Earth were created.

The Big Bang theory provides us with a comprehensive explanation of a diverse range of observed phenomena in the universe, including the discovery of the cosmic microwave background radiation, imaging of cosmic smoke at the edges of the observable universe, and the continuous expansion of the distance between galaxies, confirming that these galaxies were once closer to each other.

Thus, proponents of the Big Bang theory acknowledge the non-eternity of the universe and its beginning from a point of zero, i.e., from nothingness. They also recognize the separation of the Earth from the heavens after they were once a unified entity referred to as "Ratq". Since then, the universe has been experiencing continuous expansion and a widening gap between its galaxies. In the year 1929, the American astronomer Edwin Hubble, who worked at the Mount Wilson Observatory in California, made one of the greatest discoveries in the history of astronomy. He observed, for the first time, the continuous movement and vast separation of galaxies from one another at tremendous speeds. This discovery revealed that the universe is still expanding and has not yet reached the critical point that would lead to its collapse and subsequent implosion, indicating that it is governed by precise and intricate controls. This significant finding confirms that the components of this universe were once very close together or fused with one another at some point, thus reinforcing the theory of the Big Bang. When you imagine these galaxies moving rapidly in opposite directions, approaching each other, the closer they get, the more their mass and gravitational force intensifies. As the gravitational force increases, cohesion between the stars that make up the galaxies grows, causing the gaps between them to disappear. The gravitational pressure then increases on the stars themselves, and this process continues until all the matter comprising the universe reaches atomic size. The pressure persists indefinitely, causing the size to diminish endlessly until it becomes nothingness.

In the year 1964, Arno Penzias and Robert Wilson made a remarkable discovery about cosmic background radiation, which provided additional evidence supporting the Big Bang theory. As a result of this significant discovery and its importance in the field of physical cosmology, they were awarded the Nobel Prize in Physics in 1978. This radiation, known as electromagnetic radiation, can be detected everywhere in space within this universe. When the universe was extremely small and before the Big Bang, it was intensely hot, filled with extremely hot smoke evenly distributed throughout. The components of this smoke were hydrogen plasma, consisting of protons and free electrons due to the high temperature and energy they carried. As the universe began to expand and expand, the temperature of the plasma started to decrease until the protons could combine with electrons to form hydrogen atoms.

As the universe passed through the smoky stage that enveloped its entirety after the Big Bang, there is a single word that accurately describes that stage—smoke. This word alone represents a precise scientific expression of the reality of that phase in the universe's existence, summarizing sentences and paragraphs that scientists write today to describe that stage. In 1989, NASA (National Aeronautics and Space Administration) sent a spacecraft named "Cosmic Background Explorer" to study the cosmic background radiation from an altitude of six hundred kilometers above the Earth. The spacecraft captured images of the remnants of cosmic smoke resulting from the Big Bang process at the edges of the universe, ten billion light-years away. It confirmed that it was a smoky condition that prevailed in the universe before the creation of the heavens and the Earth.

Since explosions are usually destructive, the precise term for this process is not "explosion" but rather "Fataq" (Parting), as mentioned in the Quran. This process occurred according to precise balances and calculations, displaying utmost precision and accuracy. Thus, the force that caused this process and planned for it can only be
a tremendous force since it produced something magnificent. So why don't the discoverers of this theory wonder about what existed before the explosion? Who caused the explosion? And how could a random

explosion create such an organized universe with such extreme precision?

As for the fate of this universe, scientists expect that the speed of its expansion will slow down over time.

Mathematical calculations indicate that the expansion after the Big Bang was much faster than it is now. As the speed of expansion slows down, gravity will eventually become stronger than the force pushing matter and energy outward. This will cause galaxies and matter to move towards a hypothetical center of the universe, leading to a rapid collapse and folding in on itself. Everything in space and time will converge into an infinitely small point, returning to its original state before the Big Bang. This process is known in physics as the "Big Crunch" and is opposite to the process of the Big Bang.

What we read here is amazing and stunning by all measures.

It is extremely astonishing to read these precise scientific facts about the origin and demise of the universe, and the stages it went through, in a book that was revealed more than fourteen centuries ago, in an era dominated by ignorance, superstition, divination, and astrology! These scientific facts stated in these verses are the core of scientific theories, and even their foundation, for explaining the origin and demise of this universe! There is no doubt that the Quran's talk about matter, fusion, expansion, smoke, and folding is accurate and precise in reference and explanation! Here shouldn't we stop and ask what is the source of these verses with their amazing cosmic facts? Hasn't science today conclusively proven that the origin of the universe was in the form of a gaseous mass (smoke), which formed a connected qualitative unit with each other and fused together before being subjected to a tremendous force that led to its separation and multiplication? And here is the universe still expanding under the weight of the Big Bang.

## The Speed of Light :

There is a magnificent verse that conceals a reference to the speed of light, which scientists have only recently discovered. It is the following statement from the Quran: "And indeed, a day with your Lord is like a thousand years of what you count."

During their journey to the moon in 1969, astronauts placed glass mirrors on the lunar surface. Through these mirrors, scientists on Earth send a laser beam that reflects back to Earth. By a simple calculation, they can

## A Childish question

determine the precise distance between us and the moon. From outside the solar system, an observer sees the moon completing a full cycle around the Earth every 27.3 days. However, due to the Earth's rotation, we see the moon completing a full cycle every 29.5 days. The moon revolves around the Earth for us in a complete cycle each month, but due to the Earth's rotation in the same direction, the month appears to be 29.5 days long, while the actual time it takes for the moon is only 27.3 days. The question is, what is the distance the moon travels during its journey around the Earth in a thousand years? The idea proposed by a Muslim scientist is that the noble verse indicates two equal times, which is a type of relativity. The verse states: "And indeed, a day with your Lord is like a thousand years of what you count." If we have a day and we have a thousand years, how can we equate them? What is the common factor? Scientists consider the speed of light to be a cosmic speed that no object can practically reach. As the speed of an object increases, time slows down for it. When any object reaches this speed (the speed of light), time stops for it. This is a summary of the theory of relativity. The speed of light in a void, according to international standards, is 299,792 kilometers per second. Let's keep this number in mind because we will find it in the verse shortly. If we name the day mentioned in the verse as the cosmic day (to distinguish it from our ordinary day), we can write the following equation based on the noble verse:

## A Childish question

"And indeed, a day with your Lord is like a thousand years of what you count." Cosmic day = a thousand ordinary years. If there is a hidden relationship between the length of the day and the length of a thousand years, what is this hidden relationship that the Quran intended?

**1. Calculating the length of a thousand years:**

Since the calculation of months and years is usually based on the motion of the moon, for us, a month represents a complete cycle of the moon orbiting the Earth. As known, the moon revolves around the Earth in a monthly cycle, and after 12 cycles, a year is completed. By a simple calculation based on the actual month, we find that the moon covers a distance of approximately 21,526,112.27 kilometers around the Earth in a complete, real cycle, representing the length of its orbit during a month.

If we want to calculate what the moon covers in a year, we multiply this orbit by 12 (the number of months in a year):

21,526,112.27 x 12 = 258,313,470 kilometers. And if we want to know what the moon covers in a thousand years, we multiply the final number by a thousand: 258,313,470 x 1,000 = 258,313,470,000 kilometers.

## 2. The length of one day:

A day is approximately 24 hours, but its value in seconds according to international standards is 86,164 seconds. Now, we have the value of a thousand years, which is 258,313,470,000 kilometers, representing "distance," and we have the length of a day, which is 86,164 seconds, representing "time." To understand the hidden relationship between distance and time, we turn to the well-known equation: Speed = Distance ÷ Time. We have the known distance and the known time: Distance of the moon's orbit in a thousand years = 258,313,470,000 kilometers. Length of one day (time) = 86,164 seconds. We have only one unknown in this equation, which is the speed. By applying these numbers according to this equation, we discover the surprise: Cosmic Speed = 258,313,470,000 ÷ 86,164 = 2,997,920 kilometers per second, which is the exact speed of light. Therefore, the verse indicates a hidden reference to the speed of light by linking it to the day and the thousand years. This scientific precedence in the Quran cannot be a mere coincidence.

The truth is that the Quran contains hidden signs that no one can directly perceive. It takes hundreds of years until the appropriate era comes to reveal the miracle and serve as evidence of the Quran's truthfulness as the timeless message suitable for every time and place.

The verse has connected "day" and "thousand years," and as we have observed, the relationship between the cosmic day and the ordinary thousand years (as we count them) is a number with a value of 2,997,920, which accurately represents the speed of light. How can we explain the presence of this number in the Quran?
**Faster than the light:**

This verse in the Quran mentions: "He arranges [each] matter from the heaven to the earth; then it will ascend to Him in a Day, the extent of which is a thousand years of those which you count." Skeptics have asked whether divine affairs move at the speed of light. We say that this noble verse hints at a hidden reference to the speed of light. As for His saying, "then it will ascend to Him in a Day, the extent of which is a thousand years," this means that divine affairs ascend to the seventh heaven in one day, or a thousand years of what we count. But what does
that mean?

It means a hidden reference to the existence of a much faster speed than the speed of light!!! We know that the farthest discovered galaxy is about twenty billion light-years away, meaning that light needs twenty billion years to cross it in one day, covering a distance of 2,583,134,700,000 kilometers, and this distance is within the boundaries of the solar system. Therefore, the speed of light is not sufficient to cross the lower heaven in one day, and there must be a much faster speed than the speed of light, which some scientists believe today. Scientists have begun to notice some cosmic phenomena such as dark matter, and their belief in the existence of a much faster cosmic speed than the speed of light has grown! The idea was first proposed by two scientists in Britain and America in the late twentieth century, so say, scientists. As of the writing of this article, there are no measurements that prove this theory, but the Quran confirms the existence of a cosmic speed much higher than the speed of light. Therefore, we can say that the Quran pointed to this theory fourteen centuries ago, and this is a type of miracle. We still have a hidden reference in the verse to the shape of the path that any object takes in space, which is the curved path. All spacecraft, meteorites, and other objects move in space along a curved path and not in a straight line, due to the presence of strong gravitational fields that change the path of any object in space. Even cosmic rays, light, and other types of energy also take curved paths because they are affected by violent gravitational fields in the universe.

## Jupiter: Cosmic Guardian

This article was written in The New York Times on June 9, 2009, and has been translated with minor adjustments:

Jupiter took a bullet for us last weekend.

A body, possibly a comet that no one saw coming, crashed into the colorful cloud belt of the giant planet at some point on Sunday, scattering debris and leaving a black eye the size of the Pacific Ocean. This was the second time in 15 years that this has happened. The world watched as Comet Shoemaker-Levy 9 crashed into Jupiter in 1994, leaving Earth-sized scars that lasted for up to a year. This is Jupiter doing its cosmic job, as astronomers like to say better than we do. The story goes that part of what makes Earth such a great place to live is that Jupiter's overbearing gravity acts as a gravitational shield deflecting incoming space debris, primarily comets, away from the inner solar system where they could do to us what asteroids apparently did to dinosaurs 65 million years ago. In fact, astronomers are looking for similar formations - giant outer planets with room for smaller planets near main stars - in other planetary systems as an indicator of their hospitality to life.
Anthony Wesley, an Australian amateur astronomer who first noticed the mark on Jupiter and sounded the alarm on Sunday, praised this idea when he told The Sydney

Morning Herald: "If anything like this had hit Earth it would have been curtains for us, so we can feel very happy that Jupiter is doing its vacuum cleaner job and hoovering up all these big lumps before they come down on us." But is this warm and fuzzy image of the king of planets as our protector really true? Said Brian G., as the former director of the Central Bureau for Astronomical Telegrams, which is affiliated with the International Astronomical Union, for he spent his professional life tracking wandering bodies, especially comets, in the solar system. He also said that Jupiter is just a danger like a savior. The large planet throws many comets out of the solar system, but it also throws them inside.

Take, for example, the Lexell comet, named after the Swedish astronomer Anders Lexell. Dr. Marsden said that in 1770 it was still only a million miles from Earth, and we missed what was cosmic hair. This comet had come from outside the solar system three years earlier and passed near Jupiter, turning it into a new orbit and directly toward Earth. The comet made two passes around the sun and in 1779 passed again near Jupiter and then threw it out of the solar system again.

Dr. Marsden, who complained that the comet would never have approached Earth if Jupiter had not thrown it at us in the first place, said it was as if Jupiter had targeted us and missed. Hal Levison, an astronomer at the Southwest Research Institute in Boulder, Colorado, who studies the evolution of the solar system, said whether Jupiter is a danger or a protector depends on the source of comets. Lexell clarified, like 9 Shoemaker-Levy and perhaps the charge that just hit Jupiter, which likely came from an icy region of debris known as the Kuiper Belt, located beyond Neptune's orbit. He said Jupiter may increase our exposure to those comets.

But Jupiter helps protect us, he said, from a more dangerous group of comets that come from what is called the Oort Cloud, a deep spherical freeze surrounding the solar system a light year away from the sun. From time to time, in response to gravitational impulses from a passing star or gas cloud, a comet is launched from storage and collides inward.

The beneficial influence of the planet Jupiter comes in two forms. According to Dr. Levison, the cloud was initially populated in the early days of the solar system by the gravity of Uranus and Neptune, which swept up the debris and ejected it outward. However, Jupiter and Saturn are extremely powerful. As Dr. Levison stated, they first threw a lot of garbage out of the solar system altogether, significantly reducing the size of this cosmic arsenal. Secondly, Jupiter deflects some comets that break away from Earth and cause them to fall back, as Dr. Levison mentioned. He said, "It's a double blow against the impact." However, astronomers say that asteroids pose the greatest threat to Earth. Here, the impact of Jupiter is hardly certain. Asteroids mostly live peacefully in the asteroid belt between Mars and Jupiter, where their gravity, as the usual story goes, slightly disrupts them, preventing them from merging into a planet but potentially causing them to collide and rebound towards Earth.

This is what happened, as stated by Greg Laughlin of the University of California, Santa Cruz, to a piece of iron and nickel that traveled about 50 yards over approximately 10 million to 100 million years. The result is a mile wide, 500-foot-deep crater in the Arizona desert called the Barringer Crater, perhaps a gift from our friend and master, the planet Jupiter.

There are many aspects and arrangements that cannot occur with the precision and accuracy that cannot be wrong by even the slightest amount, but by a miraculous measure that can never be a coincidence.

A Childish question

# Chapter five

A Childish question

# Fifth Chapter

The question here is whether the universe are living beings or a pile of inanimate objects governed by random physical laws generated by infinite energies that flow through this universe without stopping. Considering the precision of its formation, this universe must be alive, governed by precise chemical, physical, and vital systems that control its overall survival and continuity. It has effects on what it contains from galaxies and the cohesion of forces. From its inception, it grows, develops, ages, and dies. Therefore, they are constantly in their place and serve their purpose. They are weak creatures with no choice but to follow their path and laws according to what is necessary for their survival. For this reason, each star or galaxy has a special role and influence that it follows and does not exceed. Simplicity overcomes them; they have bright lights and extended rays with directions and requirements. For this reason, each star has a color, weight, wind direction, quantity, how to face it, rank, location, time, and more. Also, its radius has a spiritual power that affects the entities in its psychological and mental environment. The effect of planets and stars on Earth is like the effect of drugs or elements on others without any difference in softness or density or strength or weakness or spirituality or nature.

The star has an impact according to its nature and spiritual strength. Don't you see that things without souls quickly decay and change while living things remain as long as there is a soul in them? They fix themselves on their own and push away corruption from themselves. This introduction is to facilitate understanding of the following facts...

## Is Empty Space (Nothingness) Present in the Universe?:

Based on our current understanding of space through quantum field theory, which has been experimentally proven through millions of experiments, a vacuum devoid of anything simply contradicts the laws of nature. Physics confirms that no matter how much we attempt to "suck out" all the matter in a given space, we would never succeed because elementary particles can emerge and annihilate from nothingness, where a particle and its antiparticle are created. Therefore, any vacuum is, in fact, a volcano of raging virtual particles that are constantly
being generated and dying out at lightning speed. This phenomenon is known as vacuum fluctuation. It's not surprising that this may seem incomprehensible to the human mind, as scientists themselves initially struggled with the idea that the vacuum possesses a highly variable structure and is not just "nothingness." Consequently, they devised experiments to confirm this, which are among the most precise achievements of modern science and technology. Physicists take great pride in these experiments, which measure parts of atoms' diameter, for example, to the accuracy of billions.

This precision is necessary because the effect of this vacuum fluctuation on the surrounding matter is extremely weak due to the short lifespan of the generated particles before they annihilate. Thus, direct measurement of these particles is impossible due to what is known as Heisenberg's uncertainty principle, and they can only be indirectly measured through their weak impact on the matter.

Two successful experiments have been conducted that are crucial for proving this fluctuation. The first one involves a very slight deviation in the motion of an electron around an atom, resulting from its extremely weak interaction with vacuum fluctuation. The second experiment is the Casimir effect, which took scientists approximately 50 years to reach the current level of technology to perform it. In this experiment, two parallel plates are placed in close proximity within a vacuum, and due to vacuum fluctuation, these plates are attracted to each other. Now, we know that this particular quantum fluctuation is one of the most important reasons why the laws of physics exist as they are. Without it, many cosmic phenomena, such as radiation, would not exist in this universe.

## Entropy and Life:

The thermodynamic properties of heat and life, or entropy and life, have been the subject of research since the early 20th century. In 1910, American historian Henry Adams published a thesis titled "A Letter to History Teachers," in which he proposed a theory about the history and the laws of thermodynamics and the principle of entropy. Later, in 1944, physicist Erwin Schrödinger, a Nobel laureate, addressed this field in his book "What is Life?" Schrödinger's perspective was that life depends on what he called "negentropy," but later this peculiar term was abandoned due to scientific objections, and the focus shifted to discussing what is known as free energy in this context.

Other studies have relied on Gibbs free energy in their discussions, as biological processes on Earth occur at a constant temperature and pressure in the atmosphere and ocean depths, and take a long time for each living organism.
In 1863, the physicist Rudolf Clausius published a memorandum titled "On the Concentration of Thermal Rays and Light, and on their Limits of Application," where he initially proposed a relationship based on his research and Lord Kelvin's ideas about entropy and life. Based on these ideas, the Austrian physicist Ludwig Boltzmann, in 1875, made the first conjecture about the

possibility of a connection between thermodynamics and the evolution of life. He relied on the ideas of both Clausius and Kelvin, and his thought process can be summarized as follows: "The question of the existence of life is not a struggle to obtain matter, as living organisms can acquire air, water, and soil, which are abundantly available. It is not a struggle for energy either, as energy is also abundant on Earth in the form of heat. Rather, it is a struggle against negative entropy, which can be obtained by converting energy from the 'hot sun' to the 'cold Earth'."

## Negative Entropy

In Erwin Schrödinger's book "What is Life?" published in 1944, he argues that the controller of life is not the second law of thermodynamics, meaning that living organisms do not decrease their entropy or maintain their entropy level by acquiring negative entropy. Despite the nonexistence of negative entropy, he writes in Chapter 6 of his book "What is Life?" about negative entropy: he abandons the term "negative entropy" and replaces it with the concept of free energy, which is what living organisms strive to acquire to sustain life. This sets life apart from other organizations that occur in non-living matter. Thus, it may seem that the process of life contradicts the second law of thermodynamics, which states that the entropy of an isolated system tends to naturally increase. However, there is no contradiction with this law because the principle is that entropy can increase or remain constant, but it only applies to a closed system. A closed system is characterized by no heat transfer from the outside to the inside or from the inside to the outside.

When there is heat or matter exchange between the system and its surroundings, a decrease in the entropy of the system is consistent with the second law and does not contradict it. The problem of organization and system exhibited by living organisms, where their system grows despite the second law of thermodynamics, is known as "Schrödinger's paradox" or "Schrödinger's dilemma." James Lovelock, a member of a group of scientists requested by NASA to build a device that can detect hypothetical life on Mars during planned space missions, pondered on this issue. Lovelock wondered, "How can we be sure that life on Mars will reveal itself through an examination based on Earthly life?" So, the fundamental question for Lovelock was, "What is life, and how can it be recognized?" When Lovelock discussed this with his colleagues at the Jet Propulsion Laboratory, they asked him what he would do to discover life on Mars. Lovelock replied by saying, "I will search for any decrease in entropy since it is a characteristic of life."

## Gibbs Free Energy and Biological Evolution

In recent years, the thermodynamic interpretation of evolution and its relationship to entropy has relied more on Gibbs's free energy than on entropy. This is because biological processes occur on Earth, which is characterized by a constant temperature and pressure, conditions that allow for the use of Gibbs's free energy. To express the second law of thermodynamics, Gibbs free energy is formulated as follows: reducing Gibbs free energy to a minimum is one form of the principle of minimum energy, which is based primarily on the principle of increasing entropy for closed systems. Moreover, applying the equation for Gibbs free energy can be derived in a suitable form so that it can be applied to open systems when we enter the chemical potential formula into the energy balance equation.

American biologist Albert Leningers adopts in his book, published in 1982, that the living system present in living cells during their growth and division compensates for randomness, irregularity, and disorder caused by their surroundings during their growth and division. In short, according to Leningers, "living organisms celebrate their internal systems by acquiring thermodynamic free energy from their environment, and this acquisition is done through food or sunlight; then living organisms return what they have acquired to their environment in an equal amount of energy in the form of heat and entropy." Similarly, according to what was mentioned by chemist John Avery in his book "Information Theory and Evolution" published in 2003, we find an argument indicating that the phenomenon of life, including its origin and evolution as well as human cultural evolution are interpreted against a background of thermodynamic laws, statistical mechanics and information theory. The (apparent) contradiction between the second law of thermodynamics and the high degree of complexity produced by biological systems - from Avery's point of view - finds its explanation in its informational content: Gibbs free energy that enters into biology from external sources. The process of natural selection that applies to increasing organization in a specific system can be expressed mathematically resulting from the equation of the second law in the case of open and non-equilibrium systems.

## Entropy and the Cosmic System:

The second law of thermodynamics has been expressed in various forms and formulations since its discovery. It was initially formulated by the French engineer Sadi Carnot (1829) and has been further developed by physicists such as Clausius, Lord Kelvin, and Planck.

Clausius formulated the law based on the irreversible flow of heat from a hot body to a cold one, stating that heat can't flow spontaneously from a cold body to a hot body without an external influence. Lord Kelvin focused on another expression of this law, which relates to the transformation of heat into work. His formulation stated that work cannot be obtained without the flow of heat, and this flow can only occur if there is a temperature difference.

Both Clausius and Kelvin's formulations were considered equivalent, as Clausius finally discovered a measurable physical quantity that can change in only one direction, which he called system entropy. Entropy represents the degree of randomness in physical systems and is now considered the best and easiest way to express this law. Clausius also revealed how to measure the change in entropy when a process occurs, which became more general than the formulation of the second law. It states that in an isolated system, separated from any other system, its entropy cannot decrease beyond reaching thermal equilibrium, where all parts have the same temperature. The second law indicates that change can continue in an isolated system. Two conditions are imposed on this system by the law: its total energy must not change, and its total entropy must always increase. The reverse is not possible unless effort is exerted. Hence, it is concluded that it is impossible to build a perpetual motion machine. The direction of change in such a system is irreversible, leading to chaos and increased randomness. Energy is distributed evenly over time until it reaches complete chaos and randomness. This is referred to as the heat death of the universe, where the temperature of all bodies becomes equal, and no further transfer occurs. These are the concepts of entropy and the second law of thermodynamics, which have been established as undeniable truths that reject any theory or experiment that contradicts them without discussion. Even British astronomer Arthur Eddington once said (in 1928), "I do not deny the validity of the second law, but I say that it does not apply to unstable systems where gravity plays an active role." It has become clear that there is a distinction between closed and open systems. In closed systems, entropy cannot decrease, while in open systems, entropy decreases at the expense of increased randomness in another system. This can be compared to the difference between living organisms and steam engines. Living organisms are open systems, unlike steam engines, which have a closed system. Nothing new enters or exits the suspended system, and it has been concluded that perpetual motion machines are impossible or that steam engines cannot operate at 100% efficiency due to heat loss, which we feel when we sense the heat of our laptops or TVs. On the other hand, open systems are characterized by imbalance and two opposing states of entropy, where it increases in one place and decreases in another. Thus, creating a system somewhere requires exporting entropy to another place, as we see in the world of biology and our human world, where entropy decreases with the increasing complexity of the organization, but this comes at the cost of discharging entropy outward in the form of energy.

All of this remains conditional on the existence of previous systems, whether closed or open, which initially contain low levels of entropy but increase over time according to the second law of thermodynamics. However, the situation is different when assuming frames with high entropy, as we will see... Thus, what happens in the universe as a whole is in favor of the increasing entropy, which increases everywhere and in every direction, including complex systems and some examples that can be cited. For instance, if we have a container with two types of free atoms when a molecule is formed from these two atoms, it increases the entropy because the process releases energy outward, and this entropy is added to the entire substance of the container. This is in accordance with the second law of thermodynamics, and therefore, there is no contradiction between this law and what happens in complex open systems, such as living organisms. Every increase in a system in one place comes at the cost of increasing chaos in another place. Living systems are characterized by their openness to their environment; they are not isolated, and the price they pay is the disturbance

outside. The total quantity of disturbance in the system never decreases. If the system is isolated from its environment, any change that occurs will raise the disturbance or entropy until it reaches a point where no further change occurs, and the system reaches a state of thermal equilibrium.
From a mathematical perspective, the change in entropy is equal to the quotient of the gained or lost heat by its absolute temperature when the gain or loss occurs.

## How does the cosmic system form?:

Physically, differences in energy levels according to the second law of thermodynamics allow for the possibility of forming organized structures. The flow of energy from higher to lower levels is the secret to the order in the universe, and thus the formation of life and other complex systems where entropy decreases. There is a relationship between entropy and heat or energy, as heat loss leads to a decrease in entropy and thus the formation of organized structures. Heat gain leads to an increase in entropy, as chaos spreads with random movement. This is the basic principle of thermodynamics as understood by physicists.
The truth is that heat is a double-edged sword, as it is considered an indicator of chaos when it spreads in space and when taken into account overall, but on the other hand, it leads to the formation of organized structures.

Without it, this cannot be achieved. Its determination remains dependent on the amount of this heat, as some structures require large amounts of acquired heat and energy while others do not require such massive amounts.

Therefore, just as a decrease in temperature can cause the formation of organized structures and reduce entropy relatively, as happens when water freezes, this result can also be achieved - in other cases - when the temperature rises, such as in chemical reactions where the rate of chemical reaction speed usually varies directly with the temperature according to Van't Hoff's equation. Therefore, from this perspective, an increase in temperature does not indicate an increase in entropy and chaos; hence we consider its effect relative. The second law itself is also relative since transferring heat from high degrees to low degrees does not always lead to creating an organization; rather, in certain cases, it may lead to chaos and increased entropy. A decrease in temperature to freezing point - for example - works on destroying living organisms' tissues and disrupting their functional systems; however, the overall tendency leans towards introducing chaos since even when creating organized structures there is a price paid towards outside due to leakage of wasted heat that is not utilized in forming these structures, thus becoming a source of randomness and disturbance. According to this fact, we can imagine the origin of the universe based on two different assumptions: one according to the theory of the Big Bang, and the other according to the theory of cosmic contraction. According to the first theory, physicists assume that the universe was closed and in a state of maximum trapped energy, with randomness filling this immense energy at its peak. Hence, the embarrassing question arises: how could the precise system emerge from this tremendous randomness automatically?

In his book "Emergent Science from the Autonomous System," published in 2003, the author Stephen Strogatz addressed a challenging problem regarding the existence of the autonomous system in the universe, which puzzled scientists because the laws of thermodynamics decide the opposite. It is supposed that the universe should pass through a state of extreme disturbance and deterioration towards complete randomness and chaos, while we find complete and magnificent structures all around us, such as galaxies, ecological systems, living organisms, and human beings.

Undoubtedly, this issue still poses a dilemma for physicists' thinking. According to the second law of thermodynamics, the transition progresses from order to randomness unless there are external influences or the system is open rather than closed. The prevailing assumption suggests that the universe was closed, where gravity and other forces were not yet liberated, and space was not accidental. All of this implies that the existence of the explosion was built upon sheer randomness within isolated frames, while physical theories acknowledge the necessity for the precise system to have formed from the very beginning in a unified and magnificent manner.

Without that, galaxies and life would not have been created. Until now, it has not been explained how this system formed from tremendous energy, including the highest degrees of randomness and entropy.

According to logical interpretation, entropy should increase when particles or energy are released from their trapped state. When they are released, randomness increases more and more without the possibility of returning. The situation is similar to gas particles moving randomly, leading to an increase in entropy over time continuously. Therefore, it seems almost impossible for particles and particles to move automatically from chaos to order. According to some physicists, there is a one-way direction of time that distinguishes between the past, present, and future, which is familiar to us, where things age, metals rust, and people grow old, and not the opposite. According to the second law, the movement is from order to randomness, where entropy increases, and in the long run, the latter always prevails. Any destruction of a specific system fails to restore the lost order. The second law of thermodynamics prevents the universe from returning to its initial state, which is referred to as irreversibility, where the system transforms into chaos or thermal equilibrium.

If we were to rely on the data of the Big Bang theory, the result would indicate that what happened was contrary to what is stated in the Second Law of Closed Systems, as there was a decrease in entropy from the beginning instead of an increase, which is required by the Law of Entropy. This law stipulates that the system decreases with time and entropy increases, which is reasonable for the system to transition to randomness rather than the opposite.

Mathematician Roger Penrose previously pointed out this dilemma, without being convinced of the theory of cosmic inflation. Physicists according to this theory assumed that the system formed based on expansion during the very early stages of the universe. However, in Penrose's opinion, if the beginning was chaotic and random, then the situation would remain completely chaotic and even increase as it expands according to entropy and the second law of thermodynamics. Therefore, it does not explain to us the reason for the precise system of the universe when it expands, and this is the gap as seen by this British mathematician. He concluded that according to entropy, the universe should have started with the most precise system possible, and over time, this system began to diminish according to it, meaning that it started with very small thermal deficits.

Undoubtedly, this idea contradicts the theory of cosmic and Darwinian evolution. To address this contradiction, some have suggested that the solution came from the special features of gravity. It is impossible, physically, to isolate anything from gravity, while the second law applies to isolated or closed systems. Some others have seen that cosmic expansion can create a system that was not previously present, as the universe was very hot. Then, through expansion and contraction, the universe took shape, and this thermal difference is a typical source of useful energy. For example, through contraction and expansion, hydrogen nuclei were formed, followed by light and heavy elements. In general, many physicists have resorted to considering that the universe was not completely closed, which allowed for the possibility of system solutions instead of chaotic disorder, even though their articles confirm that the universe was closed, at least before space emerged, and that a miraculous systematic event occurred due to direct expansion, related to the value of Omega.

According to the currently relied-upon inflation model, there is precision in the critical density, which is the quantity related to the speed of inflation and expansion.

## A Childish question

**This speed must be between two extremely precise limits, encompassing many stages all at once, without room for randomness or deviation in favor of either the closed universe or the open expanding universe. This limit is expressed by the value of Omega, which was finely tuned within the early moments of the universe's existence, from the Planck time to the first second. Its value at that time equals a fraction of ($10_{60}$). This tremendous precision is astonishing beyond imagination If we were to reveal the magnitude of the miracle happening in this crucial moment, we should know that the number of atoms in the entire universe is estimated to be less than this imaginary number. To illustrate the state of the miracle; if the possible speeds of the universe were numbered with each atom given a specific number, so that the sum of those numbers equals that amazing number, then the universe would have to take one speed from this huge sum, and this speed is like a specific atom that cannot be determined among the immense number of atoms. Therefore, if we fired a particle gun toward any of these atoms, the probability that the targeted atom would be hit would be equal to one out of all atoms in the universe, which is impossible. This is also true for an assumed incredible miraculous speed that would take the universe safely through both edges of tearing and great collapse.**

This is assuming the Big Bang theory. However, if we adhere to the theory of infinite space, complete openness will be achieved, followed by susceptibility to external influence and then system creation through continuous interaction between primary particles with constant motion. But with all that has been said, it is necessary to analyze closed and open-frame cases. What is stipulated by the second law assumes that there must be a previous system and does not apply to frames lacking such a system beforehand, whether closed or open. Not every closed frame leads to randomness and entropy nor does every open frame lead to organization. There is relativity based on the nature of things exposed to randomness or organization. In other words, it is entirely correct for a closed system to be subject to the entropy principle and increasing randomness over time without automatically turning into or adding more order as it is also true that an open system may undergo what helps it stay or increase organization although not necessarily always; otherwise, we would not have witnessed living organisms experiencing extinction despite living within an open system. If we assume that there is no preexisting system in closed and open frames, meaning that randomness prevails, then in this case, there may occasionally be cases of greater organization in closed frames compared to open ones. One prominent example is when we isolate several atoms within a closed box. It is expected that cases of molecular clustering will occur within the random motion of the atoms, representing a state of simple organization compared to individual atoms. On the other hand, if this quantity of atoms were to leak into the open external environment, they would disperse without forming any significant system, no matter how simple. Therefore, scientific experiments strive to isolate natural phenomena - to some extent - to discover the organized structures they give rise to, free from the external influences that may disrupt and disturb the system, such as in chemical laboratories and others.

Entropy is a term used to measure the chaos in physical systems and depends on thermal energy and temperature. It is related to the beginning of the universe, and I'm not aware if it has any relation to its end. According to the Big Bang theory, the universe was initially very small, some say the size of an atom or even smaller, and its density was extremely high. Entropy was nonexistent because everything was highly ordered and there was no chaos. Then a massive explosion occurred, initiating time, space, and everything else with it. Chaos or entropy started to increase over time, and it will continue to increase. Some refer to this as the arrow of time, as the passage of time causes an increase in chaos and entropy in the universe. I'm not aware of the relationship between this and the end of the universe. Randomness is one of the principles in the field of physics, particularly in thermodynamics and the development of statistical physics.

In classical dynamics, a system was viewed as anything that could be measured in a classical way (such as size, pressure, etc.), and measuring the randomness of a substance primarily relied on measuring changes in heat quantity. This means that it only cares about the state of the substance from the perspective of the naked eye to change its properties. This is also known as studying the effect of heat and work on the interaction between the system and its surrounding environment and recording the effect of substance properties without being concerned with what happens at the atomic level. Since a system is made up of atoms, randomness must be described more accurately. When atoms interact, their state changes rapidly at about $10_{35}$ times per second without changing it.

By looking at the state of atoms, you can get a deeper understanding of the principles and laws of thermodynamics. The science that deals with this aspect is statistical mechanics, which describes physical phenomena in terms of the statistical processing of large numbers of atoms or molecules, especially regarding energy distribution among them.

Therefore, randomness in classical dynamics is the amount of heat lost to perform work. Here you intentionally change the temperature difference either by heating or cooling. The other concept of randomness is a measure of uncertainty or randomness (measuring disturbance). For example, if you compress gas in a tube and then open this tube, what is the probability that air will come out or enter from outside?

Of course, logically speaking, the probability that air will come out from the system (tube) to its surrounding environment (air) is more likely to occur.

"The world calls this principle randomness while I see it (after understanding it) as a system rather than randomness because this principle, we called randomness, does not serve our interests " What actually happened when I opened compressed gas for air was that compressed air came out of the tube into the air; meaning energy seeks balance between tube and air; when it reached equilibrium, probability stopped who would go where, and the reason is the balance of energy between them. Energy balance means there is no change in pressure or temperature over time, so this is considered a system rather than randomness (in my opinion).

As for the end of the universe, or more accurately, our end as humans in this universe, has nothing to do with randomness.

If you are an engineer and design a system that runs on a certain amount of energy, the system will work perfectly as long as the input energy is not much higher than what is acceptable. As the controller (engineer) of the amount of energy input into the system, you can be sure that the energy does not increase or decrease.

The same goes for Earth, where the real and only source of energy that affects it as a system is the sun. As long as the incoming energy from the sun does not increase significantly, Earth will remain safe. The controller of the amount of energy from this source (the sun engineer) is God Almighty. Whenever He wants to disrupt Earth's system, He can increase or decrease the incoming solar energy. The energy used inside Earth by humans is considered insignificant for Earth.

# A Childish question

## Conclusion:

The Earth weighs approximately six trillion tons and is surrounded by a circumference of 40,000 kilometers. It rotates around its axis at a speed of 26 kilometers per minute, as well as revolves around the sun in an orbit with a length of 950,000,000 kilometers at a speed of 30 kilometers per second. It also moves with the sun in a linear path at a speed of 230 kilometers per second, all while maintaining the perfect balance that never falters. The Earth, with its seas, rivers, trees, mountains, and deserts, remains steadfast without any noticeable disturbance despite these incredible speeds. How did it become suspended in space, maintain its orbit, and achieve such precise balance without any flaws? It is preserved in its orbits, its proximity to the sun, and its stable temperature. Can it be reasonable to assume that all of this is without a planner and that this equilibrium exists without a Creator and Dominant? Is there any clearer evidence of the existence of God than this magnificent creation? Praise be to God for facilitating the gathering, comparison, and presentation of knowledge to you, dear reader, from among hundreds of scientific and philosophical sources and experiences. The essence of what has been found is closer to the truth of things, which I explain with simplicity, clarity, and indication without delving into what is difficult to understand or prove despite its reality. If you seek it, you will find it; if you believe it, you will understand it. A gesture is sufficient for the discerning person.

A Childish question

A Childish question

www.ingramcontent.com/pod-product-compliance
Lightning Source LLC
Chambersburg PA
CBHW070114230526
45472CB00004B/1254